Hayner Public Library District-Alton
0 00 30 0328506 5
P9-CQH-234

KALEIDOSCOPE

SPACE STATIONS

by
Roy A. Gallant

BENCHMARK BOOKS
MARSHALL CAVENDISH
NEW YORK

Series consultant:
Dr. Jerry LaSala, Chairman
Department of Physics
University of Southern Maine

Benchmark Books
99 White Plains Road
Tarrytown, New York 10591-9001

Library of Congress Cataloging-in-Publication Data
Space stations / by Roy A. Gallant.
p. cm. — (Kaleidoscope)
Includes bibliographical references and index.
Summary: Explains what a space station is, how it is put into orbit, and what experiments and living conditions are like there.
ISBN 0-7614-1035-X
1. Space stations—Juvenile literature. [1. Space stations.] I. Title. II. Series. III. Kaleidoscope (Tarrytown, N.Y.)
TL797.15 G35 2000 629.44'2—dc21 99-053511

Photo research by Candlepants Incorporated
Cover photo: Photo Researchers, Inc./NASA/Science Photo Library
The photographs in this book are used by permission and through the courtesy of: Photo Researchers, Inc.: NASA/Science Photo Library, 5, 10, 21, 25, 29, 40; Novosti Press Agency/Science Photo Library, 9; John Frassanito, NASA/Science Photo Library, 13; © Dick Luria/Science Source, 14; Boeing Defense and Space Group/Science Photo Library, 17; David Ducros/Science Photo Library, 22; NASA/Science Source, 26, 30; Victor Habbick Visions/Science Photo Library, 43. CORBIS: /AFP, 6; 19; © Roger Ressmeyer, 32, 34, 37, 38.

Printed in Italy

6 5 4 3 2

CONTENT

A COSMIC HOME

A space station is a small village in space. The "villagers" are astronauts who live there from several weeks at a time to more than half a year. In their space station home, they do hundreds of scientific experiments. These experiments help improve our knowledge of medicine, astronomy, and the environment. The astronauts are also trying to find out whether people will be able to live in space during long voyages of discovery.

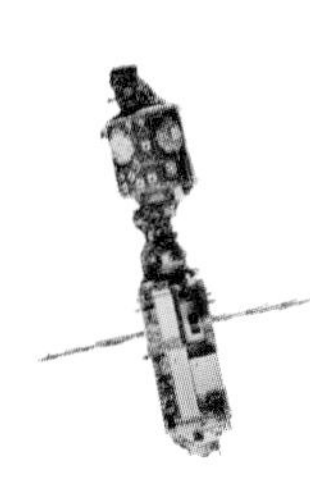

The first two pieces of the International Space Station *are the Zarya control module (center) made by the Russians and the Unity module (bottom) built by the United States.*

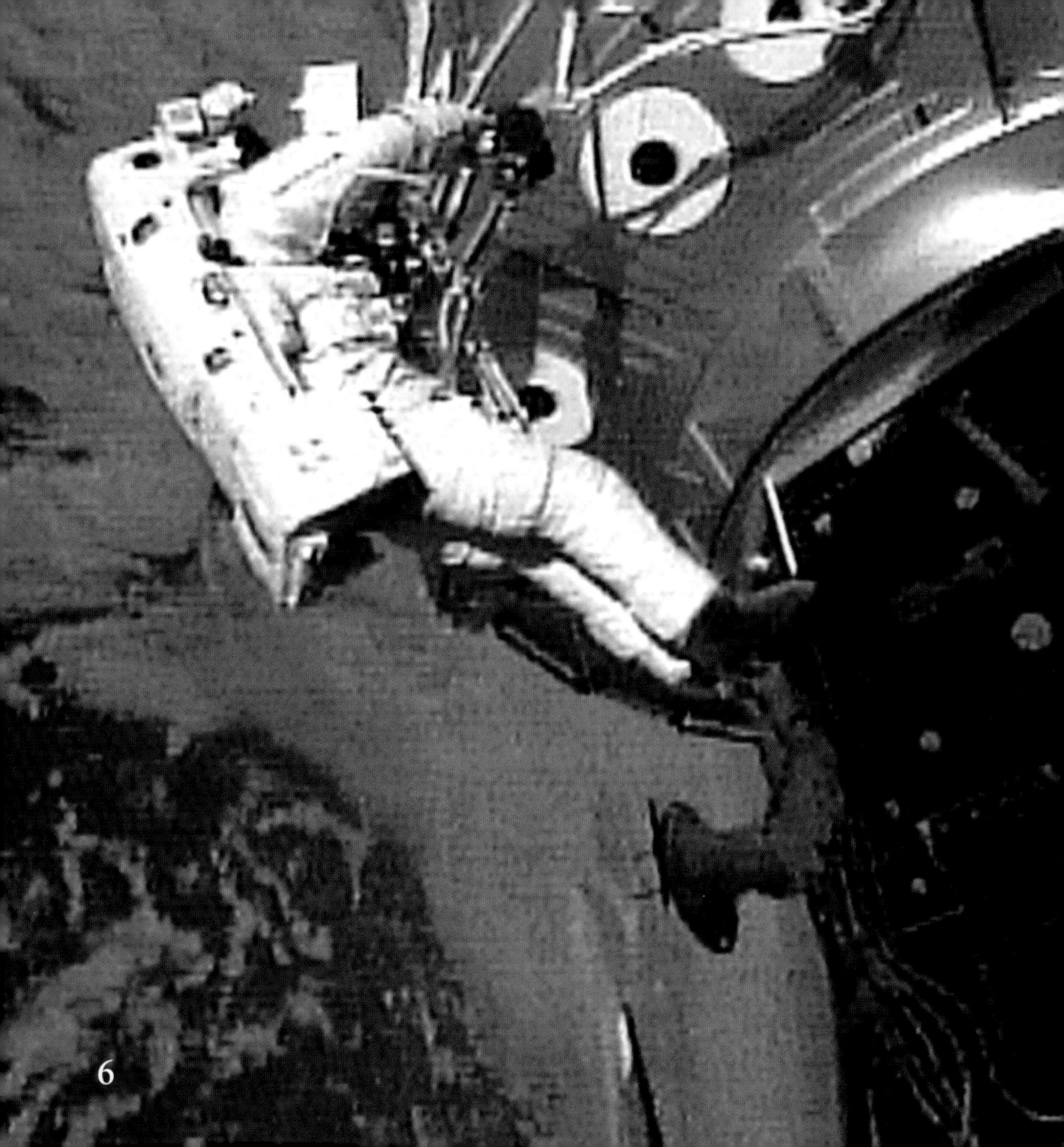

Some astronauts are scientists. Those aboard the new *International Space Station* (*ISS*) will soon be tracking hurricanes, studying Earth's climate, and checking the health of the world's oceans and freshwater. A space station gives them a better view of the planet they are trying to study. The *ISS* is now being assembled in space some 250 miles (400 kilometers) above Earth. It should be ready for its seven villagers to move into their new "home" by the year 2003. The *ISS* is just the latest in what is sure to be a long series of space stations. Let's take a look at some of the missions that have paved the way.

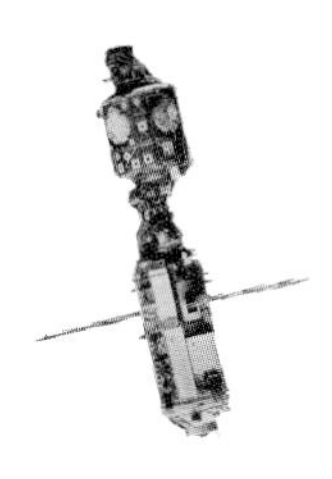

Astronaut Jim Newman works on the Unity module of the ISS *in December 1998. He and the rest of the crew then returned to Earth in the space shuttle* Endeavor.

PAST, PRESENT, FUTURE

There have been other space stations, but none as important for scientific research as the *ISS*. The first space station, *Salyut 1*, was 50 feet (15 meters) long. It was put into orbit by the Russians in 1971. Soon several more *Salyuts* were sent into space as well. Then in 1973, the United States launched its own space station. Called *Skylab*, it was about 85 feet (25 meters) long by 22 feet (7 meters) wide.

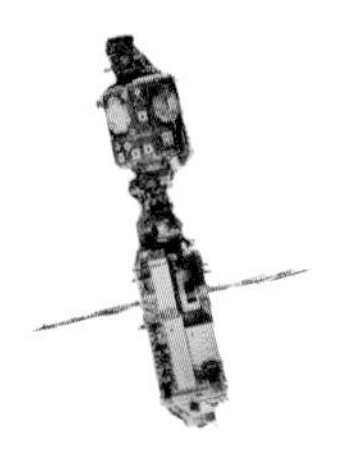

Russia's Salyut 7 *space station in orbit. On the bottom is the* Soyuz T14 *ferry spacecraft. It brought new crews to the station.*

By 1983, the United States had begun space shuttle flights. One shuttle flight carried in its cargo bay a small space station called *Spacelab.* The main purpose of both countries' early missions was to study how the human body reacts to living in space for both short and long periods of time.

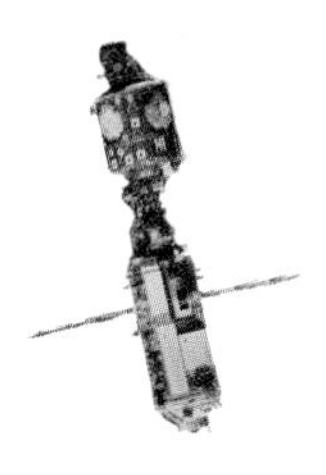

Astronaut Leroy Chiao spacewalks in the cargo bay of the Endeavor. *Below you can see Australia.*

The most important space station to date has been Russia's *Mir*. It was sent into orbit in 1986 and was home to teams of two to three astronauts who worked in the station from one to six months at a time. Having completed a mission that lasted thirteen years, in August 1999 *Mir* was shut down. The crew returned to Earth. *Mir* was going to be sent to a fiery death down through the atmosphere. But the Russians changed their minds. In 2000 crews returned to the station, and its mission continues.

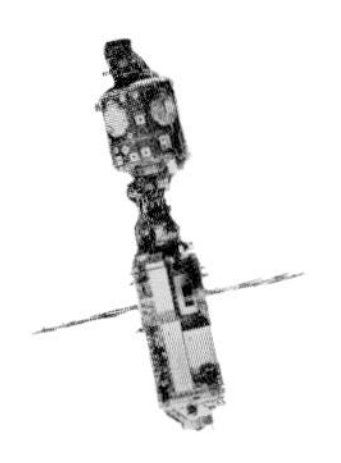

The space shuttle Atlantis *docks with Russia's* Mir *space station.* Mir *is Russian for "peace."*

Atlantis
US

Scientists must be sure the solar cells are in working order before adding them to the ISS. Cleaning and polishing them is hard work.

ASSEMBLY REQUIRED

The *ISS* is so big—355 feet (108 meters) long—has so many parts, and is so expensive that no one nation could pay the bill alone. Some sixteen countries became partners in the project. The United States and Russia are playing the largest roles, first building most of the parts and then assembling them in space. The United States is making the crew's quarters, a *laboratory* for experiments, the life-support systems, and four of the huge solar panels. The solar panels collect sunlight and turn it into the energy that powers the space station.

The Russians have built the huge living-quarters *module* plus two research modules. Both nations share the duties of running the station's navigation and guidance systems. These keep the station tilted just right to get enough energy from the Sun. They also keep the *ISS* following the right path in its orbit. In space, there is nowhere to stop and ask for directions.

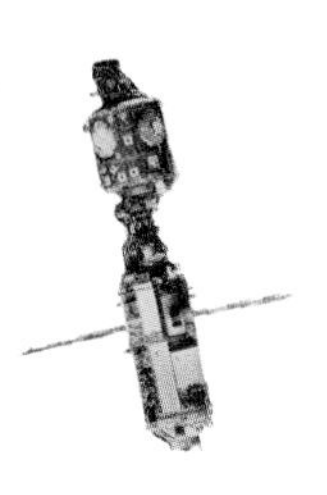

These are two modules for the ISS *made by the United States. The crew's living quarters are in front. A laboratory is in back.*

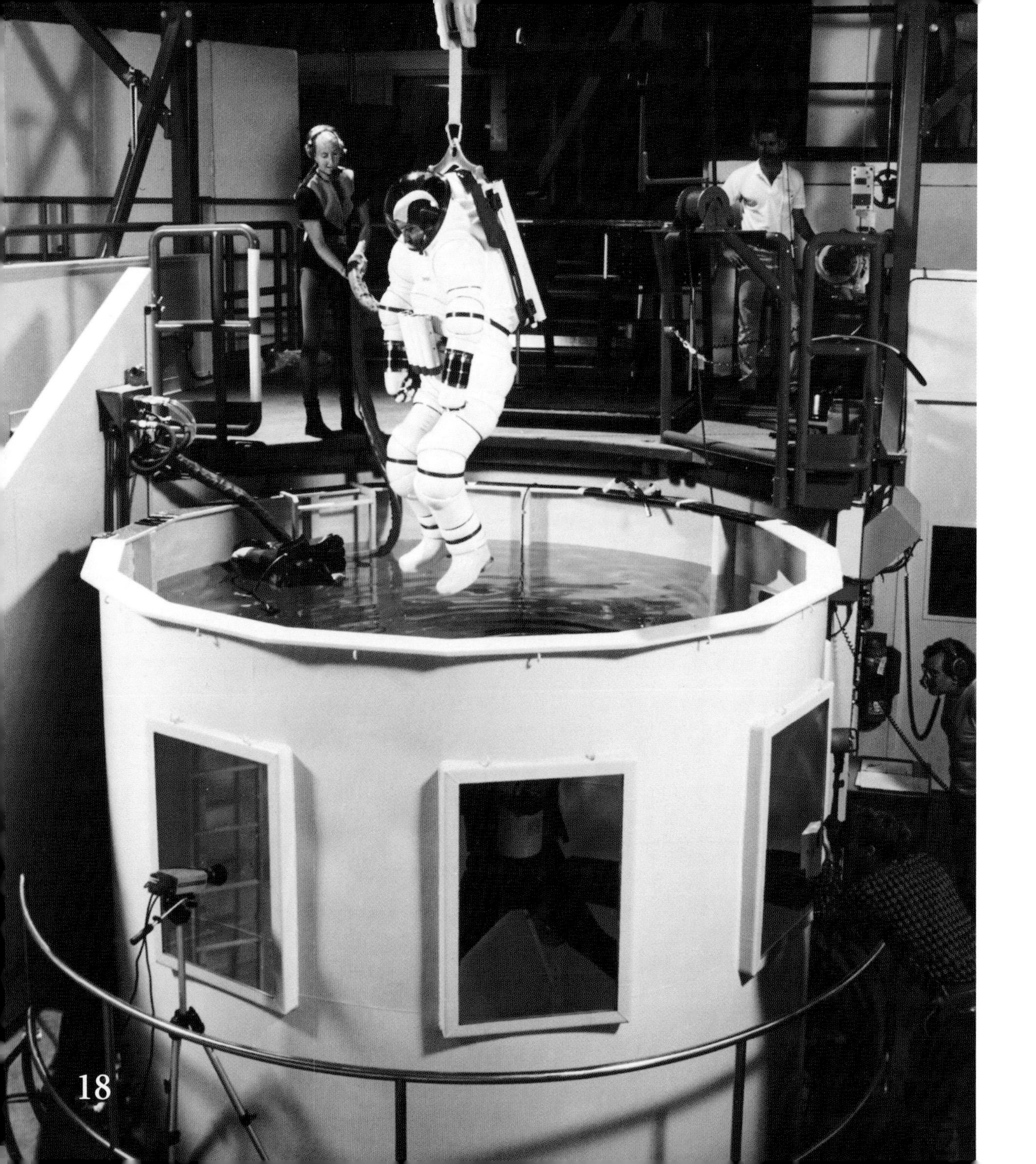

Other nations are helping as well. Canada has built a 55-foot (17-meter) robot arm that will attach to the main part of the station. The arm moves beams, panels, and other pieces into place during assembly. The European Space Agency has made a special laboratory where the air pressure is the same as on Earth. Japan is also supplying a laboratory with a platform that can be used for experiments outside the *ISS*.

The United States is also making a new type of spacesuit. It will help protect astronauts when they are working outside the station.

Early in the year 2000, astronauts began to piece the station together. Almost every month a new module was added to the station. Most were carried into orbit by a space shuttle, but several were launched by three different Russian rockets. Before the end of the year, the space station was well on its way. In all, more than 100 *ISS* pieces are being fitted together during 45 launch missions and 850 hours of space walks.

It would be impossible to assemble such a space station on the ground and then launch it into orbit. Weighing a little over 1 million pounds (453,000 kilograms), the *ISS* is simply too heavy. When finished, the station will have about a dozen linked modules.

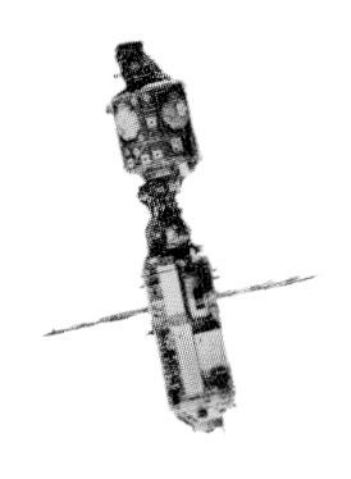

Few people have had the chance to take a space walk. Here, an ISS *worker is joining the modules Zarya and Unity.*

Air, food, and scientific tools will be brought to the ISS by this supply shuttle, the European Automated Transfer Vehicle.

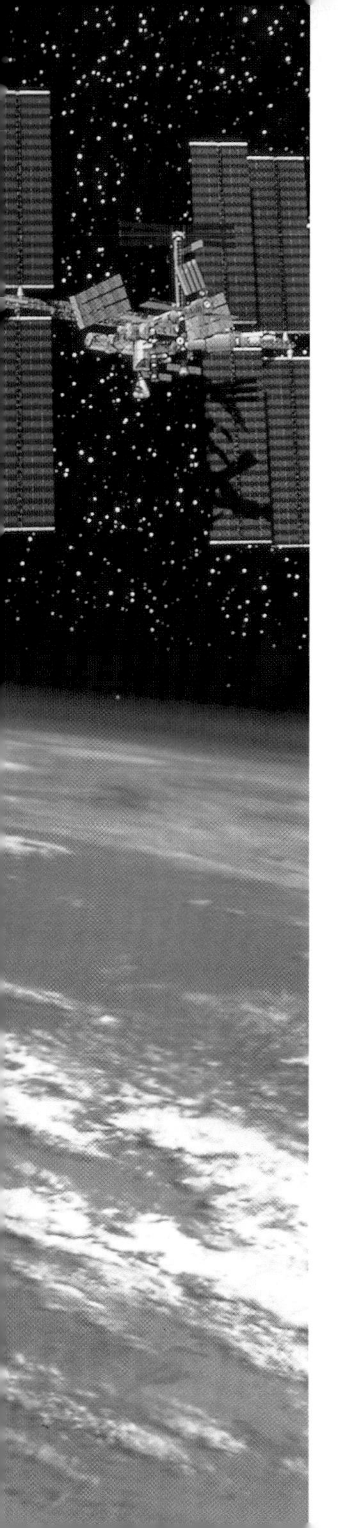

A modern research center, the *ISS* has several systems that power the station and keep it in contact with workers back on Earth. The frame, built in ten sections by the United States, is the station's backbone. Branching off it are various modules and eight solar power panels. Each is nearly two football fields in length.

Crews that shuttle workers, supplies, and equipment to the *ISS* park their spacecraft for brief visits at several docking stations. But docking can be tricky as the space station races along at 17,500 miles (28,160 kilometers) an hour. At that speed it orbits Earth once every ninety minutes.

A MIND FOR SCIENCE

What will *ISS* workers actually do up there when the station is finished? Around the clock scientists will be doing experiments to gain knowledge and improve our lives in many ways.

On a space station gravity has only one-millionth of the pull we feel on Earth. This *weightlessness* is called *microgravity,* meaning very weak gravity. Doctors want to know how weightlessness affects humans over long periods of time—how it weakens muscles, changes the way the heart, arteries, and veins work, and why bones become weaker in microgravity. Scientists aboard *ISS* are trying to find some answers.

Like astronaut Carl Walz, crew members of the ISS *will be able to swim through the air.*

Discovery
STS-41

Scientists will be using a machine called a *centrifuge* to spin around certain materials such as plant and animal tissues. The centrifuge allows them to create gravity strengths ranging from microgravity to forces stronger than gravity on Earth. They can then study the effects these different levels have on the tissues. Certain health benefits might result from being exposed to various levels of gravity. It's also important to know how humans respond to different levels of gravity if some day people are to live in space for long periods of time.

One of the jobs of the ISS *crew will be to keep records of their work. One way is to film the station's many experiments.*

Aboard the *ISS*, experiments with certain metals and ceramics may lead to new and improved materials for computers, artificial bones, satellites, and space-flight vehicles. Also, medical researchers will try to discover new ways to control or kill deadly viruses, and find out how to replace diseased tissues with healthy growing cells.

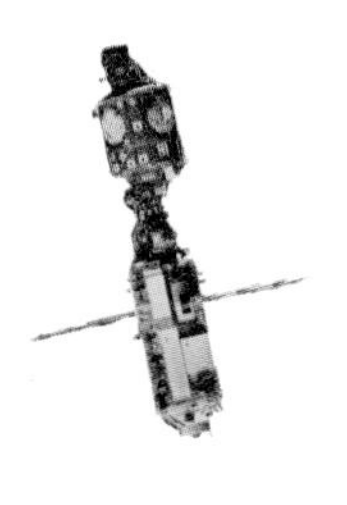

Scientists will try and learn how nerves and muscles act in microgravity. As part of an experiment aboard the space shuttle Columbia, *astronaut Kathryn Hire tosses a ball.*

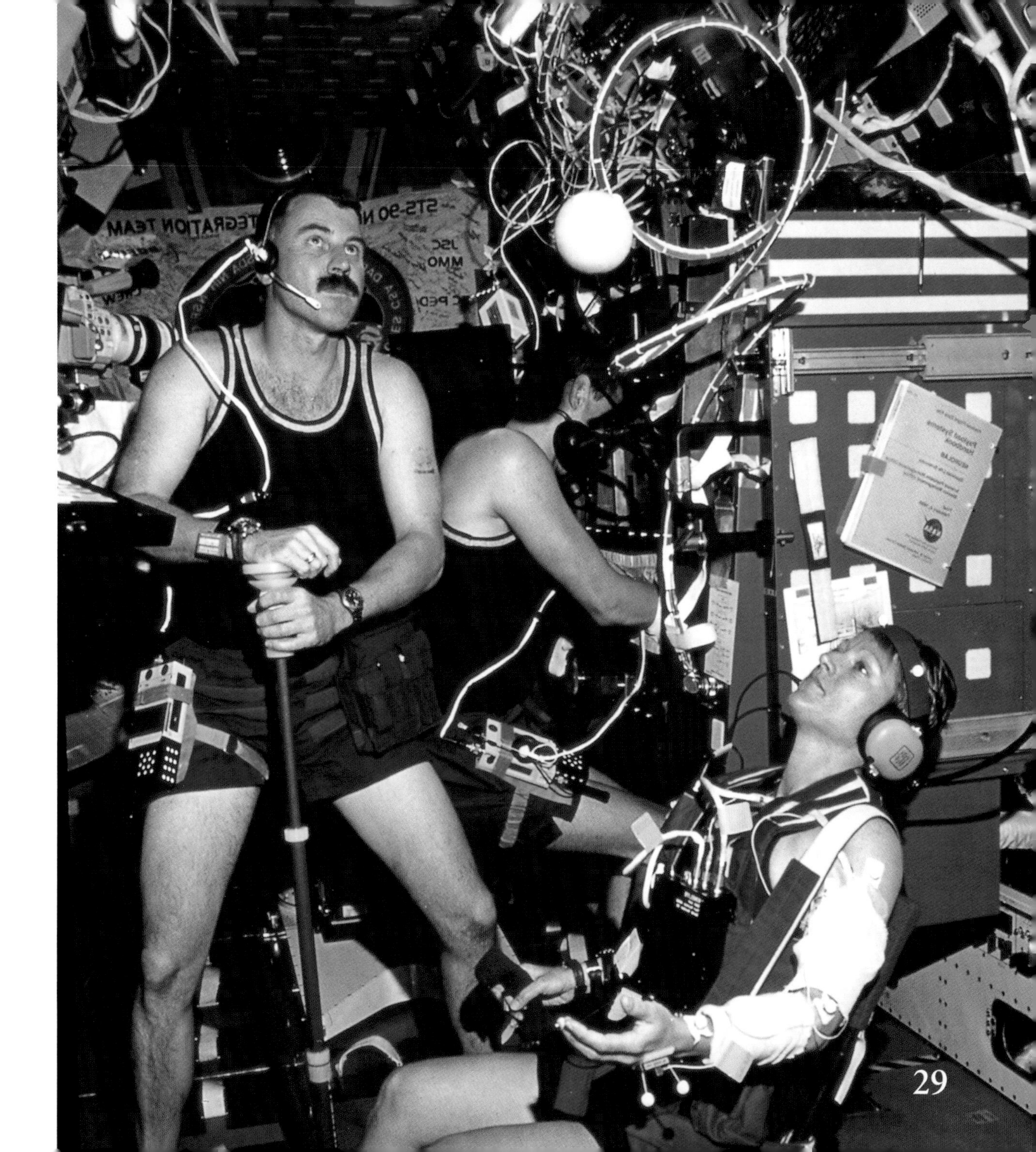

STS-90
TEGRATION TEAM
JSC
MMO

30

Instruments aboard the *ISS* will record the effects of destroying many of the world's forests and polluting its oceans and rivers. With such information, scientists will try to discover new ways to burn fuel more cleanly to reduce pollution and global warming. The knowledge that can be gained from space stations is endless.

Crystals of the gas nitrogen grown in space. They may hold the answers to a cleaner, healthier Earth.

LIFE IN SPACE

In the past, astronauts have worn spacesuits to keep them safe. *ISS* workers won't need to. This is because the mixture and pressure of the air inside the station are just right. The station is filled with the same blend of gases that we breathe on Earth —21 percent oxygen and 78 percent nitrogen. Special spongelike materials collect the carbon dioxide the station's crew exhales. The gas is then sent out into space. Heaters and circulators control the flow and temperature of the air. All wastewater is collected, cleaned, and reused.

This is a model of an ISS *module. Here, weightless astronauts are moving in three different directions.*

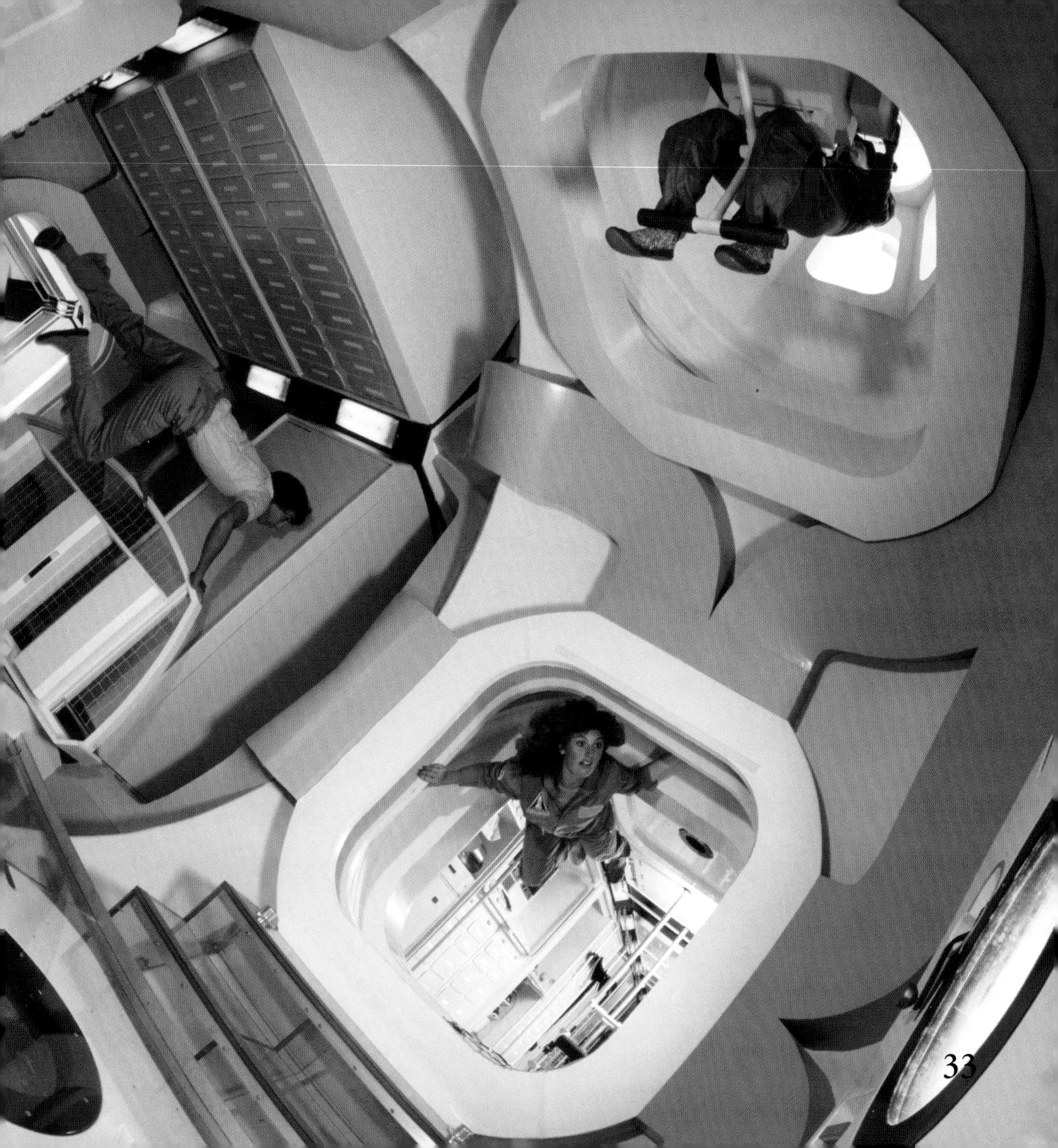

Because muscles and bones weaken in microgravity, daily exercise is important. Exercise in space takes the place of the gravity we live with on Earth. Each *ISS* worker exercises up to two hours every day on rowing machines, exercise bicycles, and a treadmill.

When it comes to mealtime, astronauts now have more choices. Space food keeps getting better, with new things to eat and new ways to prepare them. All food is tested for taste, nutrition, ease of storage, and purity.

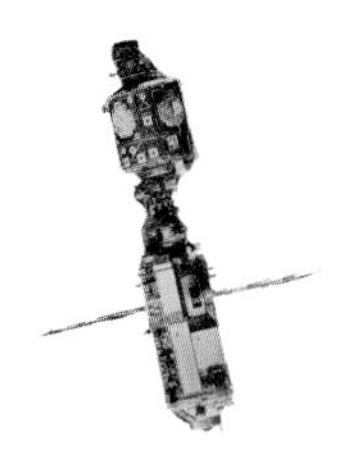

On ISS*, exercise keeps muscles strong. Here, astronaut Mario Runco Jr. works out on a rowing machine.*

Sleeping in space is easy. Just close your eyes wherever you happen to be—strapped in your seat or floating around, since in space there is no up or down. Just make sure you're attached to something so that you don't float away. Actually, the *ISS* has nooks, or small spaces, that can be closed off to create a dark, quiet place for sleeping. In microgravity there's no need for a mattress, just the straps to hold you against the padded bed board.

If you need to take a nap in space, make sure you're strapped in. Otherwise, you'll wake up somewhere else.

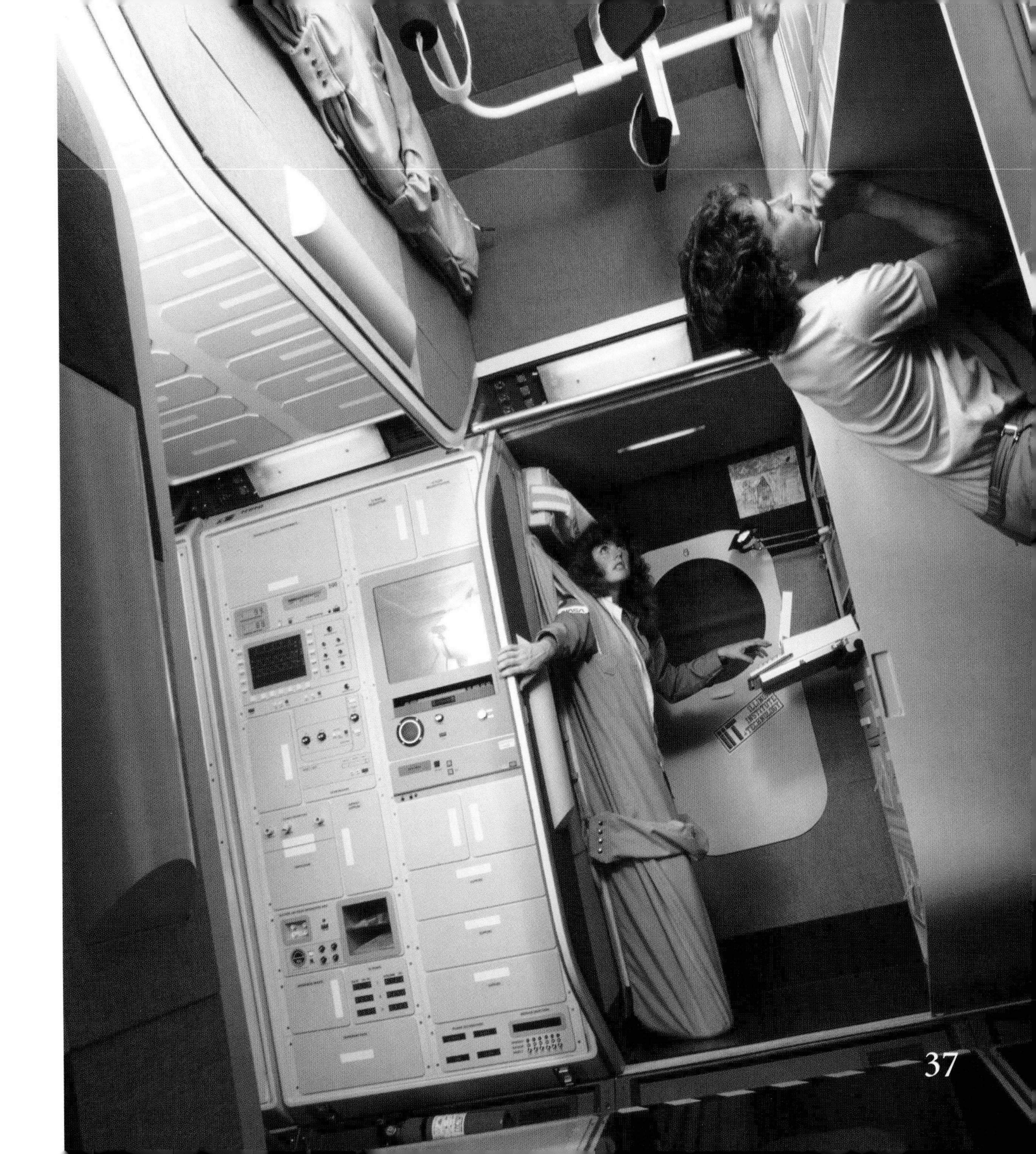

Back on Earth, workers in the control room are constantly watching the station. They are ready to help should there be a problem.

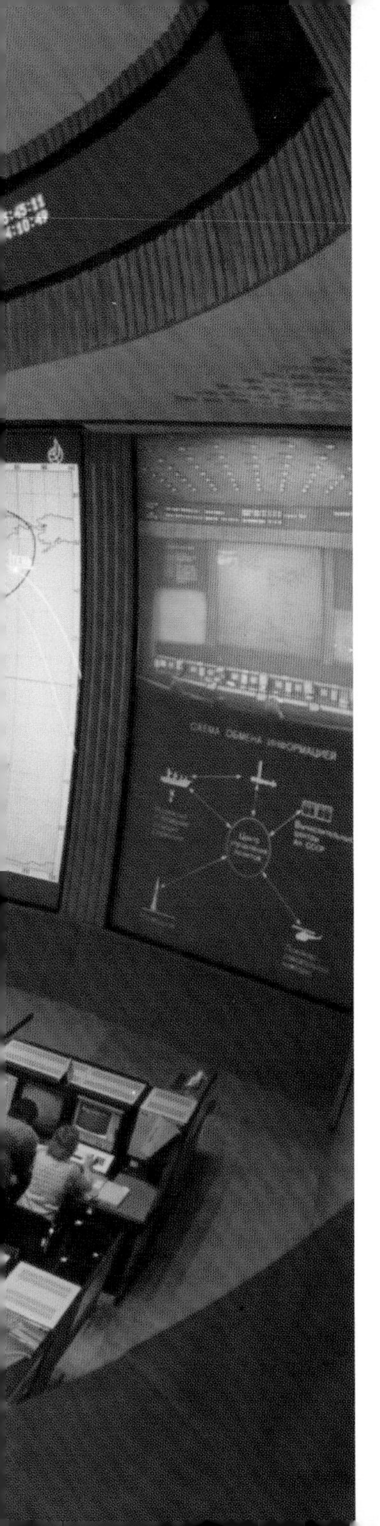

The *ISS* is designed with safety in mind. But what if something goes wrong? Two shuttles able to hold up to seven people each are docked and ready for a quick return to Earth should a problem occur. If a piece of space junk crashes into the *ISS*, or if a solar energy panel fails, the station is fixed by the crew during space walks. If the damage is serious, then special service missions from Earth would be needed. An entire module would be shut down and sealed off from the rest of the station until help arrived.

A grasping arm removes a module from the cargo bay of Endeavor *and guides it into place. The module is then joined to the main body of the ISS.*

JUST THE BEGINNING

The *ISS* is one of the most exciting space projects ever. Even so, some view it as only the second step, after *Mir,* to even grander space stations housing dozens, even hundreds, of workers. The next move may be colonies on the Moon and Mars.

But will people ever live in space, visiting their home planet only a few times during their lifetime? We must remember that we are people of planet Earth, where gravity rules our lives from the moment we are born. Our bodies are used to conditions on Earth, not to those in space. To live in space, we must bring the essential parts of our world with us. We are now learning to do just that.

The work of these scientists and astronauts has opened up a new and exciting world. While once we looked to the skies in wonder, now space is a place where some people live and work. Maybe some time in the next century students will be going to space school and reading about life on that faraway planet called Earth. Indeed, it is just the beginning.

Could this be a student catching the "school bus," reflected in the visor of her spacesuit? We'll have to wait and see.

MARS 1

GLOSSARY

centrifuge A machine that spins matter or liquid around and around very quickly to increase the effect of gravity.

International Space Station The largest space station ever, it will be finished in 2003 and hold a crew of seven. Many nations have worked together to make the *ISS*.

laboratory A place where scientists perform experiments.

microgravity The state of weightlessness when a spacecraft is orbiting Earth.

Mir A Russian space station that housed astronauts from 1986 to 1999.

module A room or part of a space station, such as a laboratory, a command center, or living quarters for the crew.

Salyut The name given to a series of several space stations launched by the Russians. In 1971, *Salyut 1* became the first station ever.

Skylab The first American space station, launched in 1973.

Spacelab A small space station launched from a space shuttle mission in 1983 to study microgravity.

weightlessness Feeling little or none of the pull of gravity.

FIND OUT MORE

Books:

Baker, David. *Factories in Space.* Vero Beach, FL: Rourke, 1987.

Barrett, Norman. *Space Machines.* Danbury, CT: Watts, 1994.

Becklake, Sue. *Traveling in Space.* Mahwah, NJ: Troll, 1991.

Berliner, Don. *Living in Space.* Minneapolis: Lerner, 1993

Bernards, Neal. *Mir Space Station.* Mankato, MN: Smart Apple, 1999.

Blackman, Steven. *Space Travel.* Danbury, CT: Watts, 1993.

Cole, Michael D. *Astronauts: Training for Space.* Springfield, NJ: Enslow, 1999.

Davis, Amanda. *Space Stations: Living and Working in Space.* New York: Rosen, 1997.

Kalman, Bobbie. *Satellites and Probes.* Minneapolis: Abdo, 1998.

Kettelkamp, Larry. *Living in Space.* New York: Morrow, 1993.

Sipiera, Diane M. *Space Stations.* Danbury, CT: Childrens Press, 1998.

Steele, Phillip. *Space Travel.* Parsippany, NJ: Silver Burdett, 1991.

Uttley, Colin. *Cities in the Sky: A Beginner's Guide to Living in Space.* Brookfield, CT: Millbrook, 1998.

Websites:

NASA *ISS* Fact Book
spaceflight.nasa.gov/station/reference/factbook/index.html

ISS Experiments
science.msfc.nasa.gov/newhome/headlines/msad13jul98_1.html

Space Station
www.maximov.com/iss/index.html

The *ISS* Era Begins!
Space.about.com/education/sciphys/space/library/weekly/aa112598.html

ISS
www1.msfc.nasa.gov/NEWMSFC/station.html

Construction Underway
cnn.com/TECH/space/2000.and.beyond/space.station/index.html

AUTHOR'S BIO

Roy A. Gallant, called "one of the deans of American science writers for children" by *School Library Journal,* is the author of more than eighty books on scientific subjects. Since 1979, he has been director of the Southworth Planetarium at the University of Southern Maine, where he holds an adjunct full professorship. He lives in Rangeley, Maine.

INDEX

Page numbers for illustrations are in boldface.